August, 1990.

To Mai-chan,

All who possess imagination, also hold
"a passport to the Land of Dreams".
Your life can be full of such wonderful
adventures and delights — "unhampered
by restrictions of time or place".

Your "kindred spirit" friend,

ポーリソ

# Spirit of Place

# Spirit of Place

## Lucy Maud Montgomery
## and Prince Edward Island

Selected and edited by Francis W.P. Bolger

Photography by Wayne Barrett and Anne MacKay

Toronto
Oxford University Press

*This book is respectfully dedicated to*
Dr Stuart Macdonald,
*son of*
Lucy Maud Montgomery

We would like to thank Dr Stuart Macdonald for his gracious permission to use his mother's writings. We would also like to thank Anne Patterson, Harry Baglole and Elizabeth Oulton for their encouragement, Christine Costello and Shirley Dillon for preparing the typescript, Ruth Campbell and her son George of 'Silver Bush' for their kindness, Jason Barrett for his patience and Roger Boulton for his editorial help.

CANADIAN CATALOGUING IN PUBLICATION DATA
Montgomery, L. M. (Lucy Maud), 1874-1942.
Spirit of place

ISBN 0-19-540389-4

1. Montgomery, L. M. (Lucy Maud), 1874-1942–
Knowledge–Prince Edward Island. 2. Prince Edward Island in literature. I. Bolger, Francis W. P., 1925–
II. Barrett, Wayne. III. MacKay, Anne. IV. Title.

PS8526.O45Z53 1982    C813'.52    C82-094210-3
PR9199.3.M6Z47

© Oxford University Press (Canadian Branch) 1982
ISBN -19-5403894
4  5  6  7  –  1  0  9  8
Printed in Hong Kong by
EVERBEST PRINTING COMPANY LIMITED

# Introduction

 LUCY MAUD MONTGOMERY was born in New London (Clifton), Prince Edward Island, on 30 November 1874, the daughter of Hugh John Montgomery and Clara Woolner Macneill. After her mother's death from tuberculosis in September 1876, and her father's decision to move to the West, it was determined that Lucy Maud would make her home in Cavendish with her maternal grandparents, Alexander and Lucy Macneill. Many years later, Lucy Maud commented on the significance of this decision: 'The incidents and environment of my childhood...had a marked influence on the development of my literary gift. A different environment would have given it a different bias. Were it not for those Cavendish years, I do not think *Anne of Green Gables* would ever have been written.'

Life with her aging grandparents was not always easy for the imaginative, sensitive, and temperamental Lucy Maud. While her grandparents provided a comfortable home, they were exceedingly strict. She was rarely allowed to attend social functions and was seldom permitted to entertain her young friends in the Macneill home. But Lucy was creative and quite ingenious, and managed to find adequate compensation through her avid love of reading, the writing of prose and poetry, and, especially, through her deep appreciation of nature.

From age six until she was fifteen, Lucy Maud attended Cavendish school, a white-washed, low-eaved building situated in a spruce grove across the road from her grandfather's gate. Characteristically, it was the natural surroundings that impressed her: 'That old spruce grove,' she related, 'with its sprinkling of maple, was a fairy realm of beauty and romance to my childish imagination. I shall always be thankful that my school was near a grove—a place with winding paths and treasure-trove of ferns and mosses and woodflowers. It was a stronger and better educative influence in my life than the lessons learned at the desk in the school-house.'

When Lucy Maud was fifteen, she moved to Prince Albert, Saskatchewan, where her father, now remarried, had established a permanent home. A desire to be with her father, and to have a respite from the exacting grandparents, motivated the move. Although she was unhappy in the West, she had her first taste of literary success while living there. A poem and short story based upon historic events associated with Cavendish beach were accepted for publication by the Charlottetown *Daily Patriot*. Her literary career, which would span some fifty years, had begun.

Lucy Maud spent only one year in Prince Albert. Homesickness for Cavendish and a strained relationship with her young stepmother more than counterbalanced her attachment to her father. In letters to Penzie Macneill, one of her close Cavendish friends, she elaborated upon her yearnings for Cavendish: 'Often in my dreams,' she wrote, 'I see the dear old shore with its brown rocks and pebbled coves and the blue waters of the sparkling gulf. . . . I long to see those dear old lanes and woodland paths again under the maples and birches, beeches and poplars. I tell you Pen, if you know when you are well off you will stick to dear old Cavendish. I've seen a good many places since I left home and I tell you I haven't seen one prettier or nicer than Cavendish, and the day on which I set foot in it once more will be the happiest of my life.' Her father eventually yielded to her constant pleadings, and allowed her to return to Cavendish in September 1891, to live again with her grandparents.

After Lucy Maud came back to Prince Edward Island, she studied for the entrance examinations to Prince of Wales College and came fifth on the Island. In 1893 she entered the college and successfully completed the program required for a teacher's license. Her first school was in Bideford, P.E.I., where she taught from 1894 to 1895. The next year, she took courses in English Literature at Dalhousie College in Halifax. She then returned to the Island, and taught at Belmont, Lot 16, from 1896 to 1897, and at Lower Bedeque from 1897 to 1898.

In March of 1898 her grandfather died. She resigned from her school immediately, and returned to Cavendish to live with her grandmother, who would otherwise have had to leave her old homestead. Apart from the eight months she worked with the Halifax *Daily Echo* as a reporter, she spent the next thirteen years assisting her grandmother in the management of the home and attached post office. Throughout all the years—whether studying, teaching, reporting, or living with her grandmother, Lucy Maud continued to write poems, short stories and serials for magazines in Canada and the United States. So successful was her writing from the time of her first publication in the Charlottetown *Daily Patriot,* that she was able to state in 1901 that she was making a 'livable income' by her pen.

It had always been her fondest hope and ambition to write a novel. Some two years after her return from Halifax in 1902, the inspiration to write her first novel came. She was looking over an old notebook and found this entry: 'Elderly couple apply to orphan asylum for a boy. By mistake a girl is sent them.' From this sketch she created *Anne of Green Gables.*

Anne Shirley proved to be such a lovable, imaginative, and captivating person that, within months of its publication in 1908, *Anne of Green Gables* became a best seller and was receiving international acclaim. Anne 'shared' Lucy Maud's love and appreciation of the pastoral beauty of Prince Edward Island. Throughout her career, the other heroines that Lucy Maud Montgomery created, Emily, Pat, Jane, Marigold, were all, like Anne, possessed of nature-loving souls.

Another memorable event of the Cavendish years was Lucy Maud's engagement in 1906 to Ewen Macdonald, the Presbyterian minister in Cavendish. The engagement was kept secret because Lucy Maud had promised her grandmother that she would always stay with her. She kept her promise faithfully, and it was not until after her grandmother's death in March 1911 that Lucy Maud could make definite plans for marriage. On 5 July 1911, at her Uncle John and Aunt Annie Campbell's home in Park Corner, Lucy Maud married Ewan Macdonald. After a two-month honeymoon in the British Isles, the Macdonalds settled in Leaskdale, Ontario, where Ewan had accepted a pastorate.

Adjustment to life outside her beloved Prince Edward Island was always difficult for Lucy Maud. Month-long holidays to the Island—in years when she could spare the time—were her great consolation. She frequently elaborated upon these visits in her letters to Ephraim Weber of Alberta and George Boyd MacMillan of Scotland with whom she corresponded regularly for some forty years. She emphasized the appeal of the Island to Ephraim Weber when discussing her 1923 vacation. 'I spent several weeks on

P.E. Island this summer and had a beautiful time.... Such fields of daisies and clover! Such sunsets and twilights and fir woods; such blue and majestic oceans and provocative, alluring landscapes....' And as she told G.B. MacMillan, the visits, as always, were inspiring. 'Then I went up to Cavendish. Delight again. Some old gladness always waits there for me and leaps into my heart as soon as I return.... A certain amount of my soul long starved mounted up on wings as of eagles. I was at home—heart and soul and mind I was at home. My years of exile had vanished. I had never been away.'

Just as in earlier years when Lucy Maud had combined a literary career with teaching and household obligations, so at Leaskdale, and later at Norval and Toronto, she managed to integrate her writing with her other responsibilities. In Cavendish, after the publication of *Anne of Green Gables*, she had produced three additional novels. At Leaskdale she published eight more books and a collection of verse. She also had three children—Chester, Hugh Alexander (who was stillborn), and Stuart.

In 1926 the Macdonalds moved from Leaskdale to another parish in Norval, Ontario. While living there, Lucy Maud published six novels. The Macdonalds lived in Norval until Ewan's failing health forced him to retire in 1935. They then moved to Toronto and purchased a new home, which they appropriately called 'Journey's End'. Lucy Maud continued to write and published three more novels.

As the 1930's drew to a close, Lucy Maud Montgomery's health deteriorated. She suffered a series of nervous breakdowns and by 1941 stimulants alone enabled her to cope with life. On 24 April 1942, at the age of sixty-seven, she died. Four days later, she was taken home to Cavendish, where she lay in state in the house that had come to be known as 'Green Gables'. She was buried in Cavendish Cemetery, in a plot she herself had chosen because 'it overlooked the spots I always loved, the pond, the shore, the sand dunes and the harbour.' Lucy Maud Montgomery had at last returned to the land of her inspiration and joy.

Francis W.P. Bolger
Charlottetown, 1981

# Acknowledgements

Quotations in the introduction and for the captions are from 'The Alpine Path', *Everywoman's World*, 1917, from letters to George Boyd MacMillan, to Penzie Macneill, and to Ephraim Weber, and from L.M. Montgomery's private, unpublished diaries. The *George Boyd MacMillan Papers* are on file at the Public Archives of Canada (MG30, D 185) as are the *Ephraim Weber Papers* (MG30, D 53). The co-operation of the Public Archives in making these papers available is appreciated. 'The Alpine Path' is in the National Library of Canada. All quotation from these sources and from L.M. Montgomery's private correspondence and diaries is by kind permission of Dr Stuart Macdonald. The quotation from *The Spirit of Canada* is by kind permission of the Canadian Pacific Railway.

Peace! You never know what peace is until you walk on the shores or in the fields or along the winding red roads of Abegweit on a summer twilight when the dew is falling and the old, old stars are peeping out and the sea keeps its nightly tryst with the little land it loves. You find your soul then . . . you realize that youth is not a vanished thing but something that dwells forever in the heart. And you look around on the dimming landscape of haunted hill and long white sand–beach and murmuring ocean, on homestead lights and old fields tilled by dead and gone generations who loved them . . . you will say, 'Why . . . I have come home!'

'PRINCE EDWARD ISLAND', *THE SPIRIT OF CANADA*, 1939

1  CAVENDISH

2 WEST RIVER

I felt like a ghost revisiting a world I had once lived in with no fellow ghosts to keep me company.

TO EPHRAIM WEBER, 19 OCTOBER 1921

3 MEADOWBANK

I had, in my vivid imagination, a passport to the geography of Fairyland. In a twinkling I could—and did—whisk myself into regions of wonderful adventures, unhampered by any restrictions of time or place.

'THE ALPINE PATH', *EVERYWOMAN'S WORLD,* AUGUST 1917

5 SOUTH SHORE

6   NEW GLASGOW

Everything was invested with a kind of fairy grace and charm, emanating from my own fancy. . .

7 EMYVALE

the trees that whispered nightly around the old house where
I slept, the woodsy nooks I explored...

the homestead fields, each individualized by some oddity of
fence or shape...

9 PARK CORNER

10   CANOE COVE

the sea whose murmur was never out of my ears—all were radiant with 'the glory and the dream'.

'THE ALPINE PATH', *EVERYWOMAN'S WORLD*, AUGUST 1917

11   NORTH LAKE

And when I went down the same old miracle took place.
The moment I set foot on the red soil it was 'home'. I had
never been away. And oh, how lovely—and lovelier—and
loveliest it was. How *satisfying*.

TO EPHRAIM WEBER, 4 DECEMBER 1927

13  HAMPTON

14 VICTORIA

15   IRISHTOWN

Shadow-waves rolling over a field of ripe wheat—all gave me 'thoughts that lay too deep for tears' and feelings which I had then no vocabulary to express.

'THE ALPINE PATH', *EVERYWOMAN'S WORLD*, AUGUST 1917

16   NORTH RUSTICO

And there was the sea. I was not prepared for the flood of
emotion which swept over me when I saw it. I was stirred to
the very deeps of my being—tears filled my eyes—I
trembled! For a moment it seemed passionately to me that I
could *never* leave it again.

TO G.B. MACMILLAN, 13 SEPTEMBER 1913

17  CAVENDISH

18   CAVENDISH

For few things am I more thankful than for the fact that I was
born and bred beside that blue St Lawrence Gulf.

'THE ALPINE PATH', *EVERYWOMAN'S WORLD*, JUNE 1917

Cavendish is a narrow farming settlement fronting on the Gulf of St Lawrence. It is about three miles long and one wide. The narrow homestead farms front on the gulf and on each one is a house.

TO G.B.MACMILLAN, 9 NOVEMBER 1904

19   MACKENZIE HOMESTEAD, CAVENDISH

20   COUSIN'S SHORE, PARK CORNER

Away to the westward, across to the harbour, the view was
bounded by New London Cape, a long sharp point running
far out to sea. In my childhood I never wearied of
speculating on what might be beyond that point—a very
realm of enchantment I felt sure. . . .

Even when I gradually grew into the understanding that beyond it was merely another reach of shore just like our own it still held a mystery and a fascination for me. I longed to stand out on that remote peak, beyond which was the land of lost sunsets.

TO G.B.MACMILLAN, 21 MAY 1909

21   COUSIN'S SHORE, PARK CORNER

We spent many afternoons on the sandshore. There's nothing in all the world like that shore. But one poetry had vanished from the gulf forever. It is never now dotted with hundreds of white sails. The fishermen now have motor boats which chug-chug out in the morning and chug-chug back at night and are not on speaking terms with romance.

TO G.B.MACMILLAN, 3 SEPTEMBER 1924

The sunset was lovely beyond words. I drank its beauty in as I walked down the old shore lane and my soul was filled with a nameless exhilaration. I seemed borne on the wings of a rapturous ecstasy into the seventh heaven. I had left the world and the cares of the world so far behind me that they seemed like a forgotten dream.

TO EPHRAIM WEBER, 10 NOVEMBER 1907

Some of our dips were taken in a heavy surf. It was the cream of bathing to stand there and let a wave break up around one's neck in a glorious smother of white foam.

TO G.B.MACMILLAN, 16 SEPTEMBER 1906

25   CAVENDISH CAPES FROM ORBY HEAD

Naturally the shore was part of my life from my earliest
consciousness. I learned to know it and love it in every
mood. The Cavendish shore is a very beautiful one; part of it
is rock shore, where the rugged red cliffs rise steeply from
the boulder-strewn coves. Part is a long, gleaming
sandshore, divided from the fields and ponds behind by a
row of rounded sand-dunes, covered by coarse sand-hill grass.

'THE ALPINE PATH', *EVERYWOMAN'S WORLD*, JULY 1917

26  CAVENDISH

There is a great *solitude* about such a shore. The *woods* are never solitary—they are full of whispering, beckoning friendly life. But the sea is a mighty soul forever moaning of some great unshareable sorrow that shuts it up into itself for eternity. You can never pierce into its great mystery—you can only wander awed and spellbound on the outer frame of it. . . .

The woods call you with a hundred voices but the sea has only one—a mighty voice that drowns your soul in its majestic music. The woods are human but the sea is of the company of the archangels.

TO EPHRAIM WEBER, 10 NOVEMBER 1907

29  CAVENDISH CAPES

But the old gulf is still the same and I saw a real storm on it once more. People there have not forgotten how to live. They have still time for sweet simple things.

TO G.B.MACMILLAN, 1 JANUARY 1933

The shore was clean-washed after the storm and not a wind stirred but there was a silver surf on, dashing on the sands in a splendid white turmoil. Oh, the glory of that far gaze across the tossing waters, which was the only restless thing in all that vast stillness and peace. It was a moment worth living through weeks of storm and stress for.

TO EPHRAIM WEBER, 10 NOVEMBER 1907

31   FRENCH RIVER TOWARDS CAVENDISH SANDSPIT

The rain of the day had ceased. The sun had come out and performed its usual miracle on those blue harbours.

TO G.B. MACMILLAN, 11 AUGUST 1929

When I go to the shore I always like to have company—the greatness, the immensity, the illimitableness of the sea turns me back on myself and I yearn for human companionship . . . .

But in the woods I like to be alone for every tree is a true old friend and every tiptoeing wind a merry comrade. If I believed seriously in the doctrine of transmigration I should think I had been a *tree* in some previous stage of existence.

TO G.B.MACMILLAN, 16 SEPTEMBER 1906

35  CAVENDISH

In a corner of the front orchard grew a beautiful young birch tree. I named it 'The White Lady', and had a fancy about it to the effect that it was the beloved of all the dark spruces near, and that they were rivals for her love. It was the whitest straightest thing ever seen, young and fair and maiden-like.

'THE ALPINE PATH', *EVERYWOMAN'S WORLD*, JULY 1917

37   WHITE BIRCH

. . . we played house and gardened and swung and pick-nicked and climbed trees. How we did love trees! I am grateful that my childhood was spent in a spot where there were many trees, trees of personality, planted and tended by hands long dead, bound up with everything of joy or sorrow that visited our lives. When I have 'lived with' a tree for many years it seems to me like a beloved human companion.

'THE ALPINE PATH', *EVERYWOMAN'S WORLD*, JULY 1917

I'm going back to the woods now to get some ferns. Want to come? I could make you useful opening gates and chasing cows. There are always cows back there and I'm horribly scared of them!

TO G.B.MACMILLAN, 23 AUGUST 1905

The woods are *getting ready* to sleep—they are not *yet* asleep
but they are disrobing and are having all sorts of little
bed-time conferences and whisperings and good-nights . . . .

I can more nearly expect to come face to face with a dryad at this time of the year than any other. They are lurking behind every tree trunk—a dozen times I wheeled sharply around convinced that if I could only turn quick enough I could catch one peeping after me. Oh, keep your great vast prairies where never a wood-nymph could hide. I am content with my bosky lanes and the purple peopled shadows under my firs.

TO EPHRAIM WEBER, 10 NOVEMBER 1907

41  *PHOLIOTA (ALBOCRENULATA)*

43    BALSAM HOLLOW TRAIL, GREEN GABLES, CAVENDISH

A brook was laughing to itself in the hollow. Brooks are always in good spirits. They never do anything but laugh. It is infectious to hear them, those gay vagabonds of the valleys.

TO G.B. MACMILLAN, 9 NOVEMBER 1904

And all the old beauty. I had forgotten that the ponds and rivers of P.E. Island were so brilliantly, so unbelievably blue....

LAKE OF SHINING WATERS, PARK CORNER

I walked and prowled by night and day as I had not done
for years.

TO G.B.MACMILLAN, 27 DECEMBER 1936

46   GUERNSEY COVE

It is the sea which makes Prince Edward Island in more senses than the geographical. You cannot get away from the sea down there. Save for a few places in the interior, it is ever visible somewhere, if only in a tiny blue gap between distant hills, or a turquoise gleam through the dark boughs of spruce fringing an estuary. Great is our love for it, its tang gets into our blood: its siren call rings in our ears, and no matter where we wander in lands afar, the murmur of its waves ever summons us back in our dreams to the homeland.

'THE ALPINE PATH', *EVERYWOMAN'S WORLD*, JUNE 1917

Much of the beauty of the island is due to the vivid colour contrasts—the rich red of the winding roads, the brilliant emerald of the uplands and meadows, the glowing sapphire of the encircling sea.

'THE ALPINE PATH', *EVERYWOMAN'S WORLD*, JUNE 1917

50 'SILVER BUSH', CAMPBELL HOMESTEAD, PARK CORNER

I came over to Park Corner, where I am at present staying with an aunt. It is a beautiful spot here—a big farmhouse surrounded by orchards and beech woods with a splendid pond before the door.

TO G.B. MACMILLAN, 4 MAY 1911

'The Lake of Shining Waters' is generally supposed to be Cavendish Pond. This is not so. The pond I had in mind is the one at Park Corner, below Uncle John Campbell's home.

'THE ALPINE PATH', *EVERYWOMAN'S WORLD*, SEPTEMBER 1917

51   LAKE OF SHINING WATERS, PARK CORNER

52  IRISHTOWN TO BURLINGTON ROAD

53  CAPE TRYON LIGHT

The Park Corner jaunts were always delightful. To begin with, it was such a pretty drive...

54 CAVENDISH CAPES

those winding thirteen miles through hill and wood,
and river and shore.

'THE ALPINE PATH', *EVERYWOMAN'S WORLD,* JULY 1917

We drove along red roads with daisies blooming on their banks, past little hollows full of scented fern, past old stone dykes festooned with wild strawberries over looping blue rivers.

TO G.B.MACMILLAN, 6 FEBRUARY 1928

57 GUERNSEY COVE

58    CLOVER

To me the common red clover of our hayfields always seem masculine. The fine white or faintly tinged pink clover is a little lady but the big chubby red clovers are sturdy country lads.

TO EPHRAIM WEBER, 8 APRIL 1906

Next to us was a vacant lot full of daisies—the one where
was taken the snap I sent you . . . .

60   DAISIES

It was a place of haunted loveliness in the twilights—
and over the river were daisied fields as white as snow.

TO G.B. MACMILLAN, 3 SEPTEMBER 1924

She is gone forever from that big, beautiful, orchard-embowered house that was the wonder castle of my childhood where Aunt Annie reigned as queen, dispensing the gracious and lavish hospitality for which she was forever famous.

TO EPHRAIM WEBER, 1 NOVEMBER 1924

62   'SILVER BUSH', CAMPBELL HOMESTEAD, PARK CORNER

My first week was spent with an old college chum in her summer bungalow on the south side, built just where the North River empties into Hillsborough bay. . . .

63   MOUTH OF NORTH RIVER

It was situated between two range lights that burned enchantingly through the twilights, pearl white against the ethereal skies. Down the harbour there were more range lights and far out, seemingly in mid-harbour shone the far away lighthouse on Point Prim—a beacon in 'fairylands forlorn'.

TO G.B. MACMILLAN, 3 SEPTEMBER 1924

66   CROW BUSH COVE PROVINCIAL PARK

As I came home in the afterlight I saw a sight that filled me with rapture. To my right was a cluster of tall, gently waving spruces. Seen in daylight, these spruces are old and uncomely—dead almost to the top, with withered branches. But seen in that enchanted light, against a sky that began by being rosy saffron and continued to be silver green, and ended finally in crystal blue, they were like dark, slender witch-maidens weaving their spells of magic in a rune of elder days. How I longed to share in their gramarye—to have fellowship in their twilight sorceries.

TO G.B. MACMILLAN, 21 MAY 1909

68 (a)    RED FOX, BASIN HEAD

68 (b)    SNOWSHOE HARE

69   RASPBERRIES

We picked berries in the wild lands and fields back of the
woods, going to them through wooded lanes fragrant with
June bells, threaded with sunshine and shadow, carpeted
with mosses, where we saw foxes and rabbits in their native
haunts.

'THE ALPINE PATH', *EVERYWOMAN'S WORLD*, JULY 1917

It's a pearl of a day and the old hill road is lovely. Just now it is scarfed with a ribbon of golden rod and airy smoke blue asters—so beautiful and yet suggestive of sadness in that they are forerunners of autumn. In fact the country people hereabouts call asters by the pretty poetic name of 'farewell summers'.

TO G.B. MACMILLAN, 31 AUGUST 1908

70   ASTERS AND GOLDEN ROD, STANLEY BRIDGE

71   LEAVES, KLONDIKE ROAD

. . . amid all the green a scarlet leaf burned as if
Autumn had walked through and pushed a branch aside here
and there leaving the stain of her fingers where they touched.
The woods always seem to me to have a delicate, subtle life
all their own that epitomizes the very spirit of all the seasons
in turn and is never out of harmony with the time o' year.

TO G.B.MACMILLAN, 16 SEPTEMBER 1906

72  HAMPTON

The air was very frosty and clear. There were wonderful lakes of crimson and gold among the dark western hills.

TO G.B. MACMILLAN, 9 NOVEMBER 1904

74  IRISHTOWN

It has been awfully cold today and I've gone about with my teeth chattering. But we have had the loveliest fall, ever since the first of September, all purple and gold and mellowness in earth, air and sky.

TO EPHRAIM WEBER, 8 OCTOBER 1906

76 SPRINGBROOK

It's a cold frosty night out with a skim of snow on the ground and a silvery. . . moon floating over the orchard in a sea of saffron yellow sky.

TO G.B. MACMILLAN, 9 NOVEMBER 1904

A Canadian winter has a wonderful zest and sparkle and even the storms have a wild white majesty of their own. A still winter evening among the wooded hills with the sunset kindling great fires in the westering valleys is something I would not exchange for a month of hot July days.

TO G.B.MACMILLAN, 31 AUGUST 1908

We have had a perfectly awful winter here. The like has never been known even by that mythical personage the oldest inhabitant. We thought last winter terrible but it was not as bad as this. We have had nothing but storm after storm, train blockages, and irregular mails.

TO EPHRAIM WEBER, 7 MARCH 1905

80  EMYVALE

Tonight coming home from a tramp over snowy hills I halted a moment to look over the orchard fence at my flower bed. Not that I could see much of it—it is heaped over with snow drifts fifteen feet deep, gleaming in the twilight like a mausoleum of marble built over buried dreams.

TO EPHRAIM WEBER, 7 MARCH 1905

81 SNOW SCENE

I am very homesick and feel as if I would exchange all the kingdoms of the world and the glory thereof for a sunset ramble in Lover's Lane.

TO G.B.MACMILLAN, 20 JANUARY 1912

One evening I spent wandering about the graveyard where so many of our dead people lie. It was not a sorrowful tryst. I felt very happy and among friendly presences. I felt again acutely the peculiar, indefinable charm of P.E. Island. A certain wellspring of fancy which I thought had gone dry in me bubbled up as freshly as of old.

TO G.B. MACMILLAN, 26 FEBRUARY 1919

A certain amount of my soul long starved mounted up on wings as of eagles. I was at home—heart and soul and mind I was at home. My years of exile had vanished. I had never been away.

TO G.B. MACMILLAN, 3 SEPTEMBER 1924

85   COUSIN'S SHORE, PARK CORNER

86    SPRINGBROOK